GW01326132

DEREK JARMAN

My Garden's Boundaries are the Horizon

DEREK JARMAN

My Garden's Boundaries are the Horizon

GARDEN MUSEUM

Derek Jarman *My Garden's Boundaries are the Horizon*

Images reproduced with permission of
© Keith Collins Will Trust, Amanda Wilkinson Gallery,
London, Howard Sooley and Edina van der Wyck
Super 8mm images: courtesy of LUMA Foundation
The Garden stills: courtesy of Basilisk Communications
Portrait of Paul Reynolds (during the making of Angelic
Conversation): courtesy of James Mackay
Pages 20/21 and 22/23 courtesy of Thames & Hudson Ltd.
Additional picture research by Christine Lai

Exhibition curated by Emma House

First published in 2020
by the Garden Museum
5 Lambeth Palace Road
London SE1 7LB
020 7401 8865

www.gardenmuseum.org.uk

The Garden Museum is a registered
charity, number 1088221

ISBN: 978-1-5272-5916-4

Design by Webb & Webb Design Limited
Printed in the UK

This exhibition and catalogue is generously
supported by The Linbury Trust

Previous page: Howard Sooley, *Derek Jarman at Prospect Cottage*

CONTENTS

1956. RAF Northwood

INTRODUCTION

CHRISTOPHER WOODWARD
DIRECTOR, GARDEN MUSEUM

When Derek Jarman came to live on the bare shingle of Dungeness in 1987, he carried a lifetime of gardens within him. Writing *Modern Nature* in the fisherman's shack, 'the spring flowers' of his first home in the suburbs of London were his 'earliest memories'. His next childhood garden was Villa Zuassa on the shore of Lake Maggiore, a residence for military families; his father was a senior commander in the Mediterranean Allied Air Forces. There were 'old roses trailing into the lake' and 'abandoned avenues of mighty camellias'. His parents' present for his fourth birthday was *Beautiful Flowers and How to Grow Them*, an illustrated manual of 1926. Next, an apartment in Rome, and picking pine kernels in the Borghese Gardens. In England, a last idyll in a succession of billets was Curry Mallet, a 16th-century manor house near Taunton which creaked with Elizabethan mystery. The red valerian which grew wild on Dungeness had grown in the cracks of its stone garden walls.

The beauty of such Edens was intensified by knowledge of their loss. It was in Rome that Jarman saw his first film, walking past ruins and widows to a matinee of *The Wizard of Oz* (1939). An usherette found him hiding under the seat. In appearance, Prospect Cottage is a kind of twin to Dorothy's house; when the wind blows on Dungeness it feels like it might be picked up. And after purchasing the Cottage Jarman and his new partner Keith Collins stripped out curtains and carpets to expose its bare wooden bones.

Derek Jarman in his parents garden at RAF Northwood, London, 1956

Derek with his mother Betts and sister Gaye, Manor Cottage, 1945

Jarman bought the cottage for £32,000 in May 1987 from the £100,000 in the will of his father, who had died at the end of 1986. It was on 22nd December that - as Tony Peake describes in his outstanding biography, *Derek Jarman* (Little, Brown, 1989) - Jarman was diagnosed with HIV. At lunch immediately after he declared to a friend 'I've got to get as much out of life as possible'. But in 'the grey winter dawn', sleepless, he fantasised about another season in which a blackbird sang in the flowerpots stacked on the staircase at Phoenix House, the flats on Shaftesbury Avenue. And he wrote in his diary, never published:

> 'We must fight the fears that threaten our garden, for make no mistake ours is the garden of the poets of Will Shakespeare's sonnets, of Marlowe, Catullus, of Plato and Wilde, all those who have worked and suffered to keep it watered'.

Jarman's garden is the garden of fairy tales: a metaphor of enchantment, and love, and fear. Fear was 'poison ivy'. When he bought Prospect Cottage it was on a whim. Filming Tilda Swinton chasing the lighthouse beam on a shingle formation he saw as a 'fifth continent' a 'For Sale' sign was in the window.

'The shingles preclude a garden', he wrote in his diary on 18th August 1987. There was no soil between the stones, wind, and scorching summers. But Prospect Cottage was not, as is sometimes assumed, a garden of beachcombed stones and sculptures and 'right plant, right place'. It is a wilful fusion of Jarman's fantasies with that environment. On Monday 9th January 1989, as recorded in *Modern Nature*, he bought and planted thirty roses from Rassells of Kensington, a favourite nursery where a friend at the till 'laughed at the idea'. *Rosa mundi* was the crusader's rose of *Roman de la Rose*, he continues, and still on the shelves of Prospect Cottage are books which show his horticultural knowledge - such as Peter Beales on roses - but also his romantic fascination with the folklore and meanings of flowers. The plans on pp. 21 - 22 recorded what he had planted, as he 'detested' plastic labels, Keith Collins remembered - and that 'Derek imagined himself surrounded by a forest of impenetrable thorns, eventually hacked down by a true-hearted, handsome prince' (quoted in *Derek Jarman's Sketchbooks*, Thames & Hudson, p.197). Three roses caught Christopher Lloyd's attention the next summer. That year, Howard Sooley became a friend and, in a memoir of gardening together we are proud to publish, the garden took shape as both an adaptation of existing vegetation and continued experiments in plant purchases achieved by a lot of carrying and digging in bags of earth, and replacing stones on top'.

At boarding school Jarman won prizes for his garden plot; a family album shows the rockery made as a teenager in that back garden in Northwood. In his twenties, a star shot out of The Slade School of Art, he designed an abstraction of an 18th-century formal garden for the set of *Don Giovanni*, the ENO's 1968 debut at The Coliseum; for the film *The Devils* two years later a second garden of abstract topiary. When his sister Gaye converted a country house in Herefordshire, The Verzons, into a proto-gastropub he designed a garden of geometric evergreens. That love for the Elizabethan gardens is expressed in *The Angelic Conversation* (1985), shot at Montacute, the National Trust property in Somerset, to the narration of Shakespeare's sonnets. 'He knew exactly which garden he wanted to film in' remembers producer James Mackay.

Gardens flicker through the stills James has chosen from their collaborations (pp. 14-19); the greenest are the pot plants in *Studio Bankside*, a first Super 8, shot in one of the sequence of ruinous studios in which he lived above the rusty Thames shore. In one, a glass greenhouse was erected as a bedroom.

To Tony Peake, the purchase of Prospect also reflected a sudden possibility of permanence with Keith Collins, whom he met earlier in 1986. Its preservation by Collins for more than twenty years is an act of great love. But a garden cannot be preserved as exactly as a studio of congealed paints, or a shelf of books. It is exposed to the dynamics of a wild nature and, as importantly, when you garden the inner pulse, the digging and sweating, quickly takes over from the external image.

'That blasted book' said Collins of the iconic *Derek Jarman's Garden* (1995), a collaboration between Jarman, Sooley, and Thomas

The Devils Louis XIII Garden Scene, 1970 (Photograph Louis Haugh)

A Garden for the Verzons, 1973 (Photograph Louis Haugh)

Neurath of Thames & Hudson when I interviewed him on this theme a decade ago. 'I feel imprisoned by the photographs'. At Shipley's, the art bookshop on Charing Cross Road, he'd overheard a voice: '"Oh, I was there last week. All the plants are dead". It was mid- November'. But suggest that the vision should crumble into a more intangible but unchanging afterlife of images and memory and 'That destructiveness is so 1977'.

This exhibition coincides with The Art Fund and Creative Folkestone's campaign to preserve the Cottage as an artist's retreat. But the garden is not a site-specific art assemblage, as, I worry, it is seen by the international art collectors circling in fur coats on my last visit. It was, as Howard describes, an adventure paused at a moment in time; to keep it alive, we must go back to understand Jarman's ideas, failures, and intents.

On 24th June 1990 Christopher Lloyd took Beth Chatto on a picnic to the beach, recorded in our archive by a photograph on the reverse of which she has written in biro 'The day Christopher wandered into Derek Jarman's garden at Dungeness'. As he remembered (pp. 67-74) walking back to the car 'I spotted some brilliant flower colour; I headed for it... It hadn't just happened'. Jarman stepped out, and the three most influential gardeners in Britain had tea together. Visits, and plant lists, followed, and Lloyd wrote the first analysis of a garden which combined 'total originality' with 'the very English garden tradition' of growing plants to a place and, equally, of ignoring advice as to what is 'impossible'.

Interviewed by Lloyd's friend Pavord for *The Independent* on Sunday (pp. 76-79) Jarman was joyous: 'Every flower is a triumph. I've had more fun from this place than I've had with anything else in my life. I should have been a gardener'.

And the place has changed for ever as the plants he collected, or bought and mulched, have formed a compost between the dry stones. For the first time, Sooley notes, grass grows on the ness.

FILM STILLS

Film stills selected by Derek Jarman's Producer James Mackay author of *Derek Jarman Super 8* (2014). *The Garden* was filmed on location at Prospect Cottage.

GARDEN JOURNALS

Throughout his career Jarman collated his working notes in books purchased in Venice, later adding decorative gilding to their covers. During 1989 and 1990 he worked on five notebooks dedicated to his garden. They include his plan recording his planting scheme, as he did not like plant labels and roses bought at Rassells of Kensington in January 1989. They also record his daily routines, work and friendships. These books formed the basis for his gardening memoir *Modern Nature*.

ARTWORKS

Jarman's move to Prospect Cottage led to several new series of artworks. For the first time in many years he started to reintroduce three-dimensional items into his works incorporating found objects from the beach. In January 1990 he saw a landscape by Frank Auerbach which led him to return to a pure form of landscape painting. The resulting views of Dungeness and the local marshland are rich in colour with thickly applied impasto.

A Journey to Avebury, 1971-72

Fire Island, 1974

The Garden, 1990

The Garden, 1990

Studio Bankside, 1970

Opposite: *Angelic Conversations, 1985*

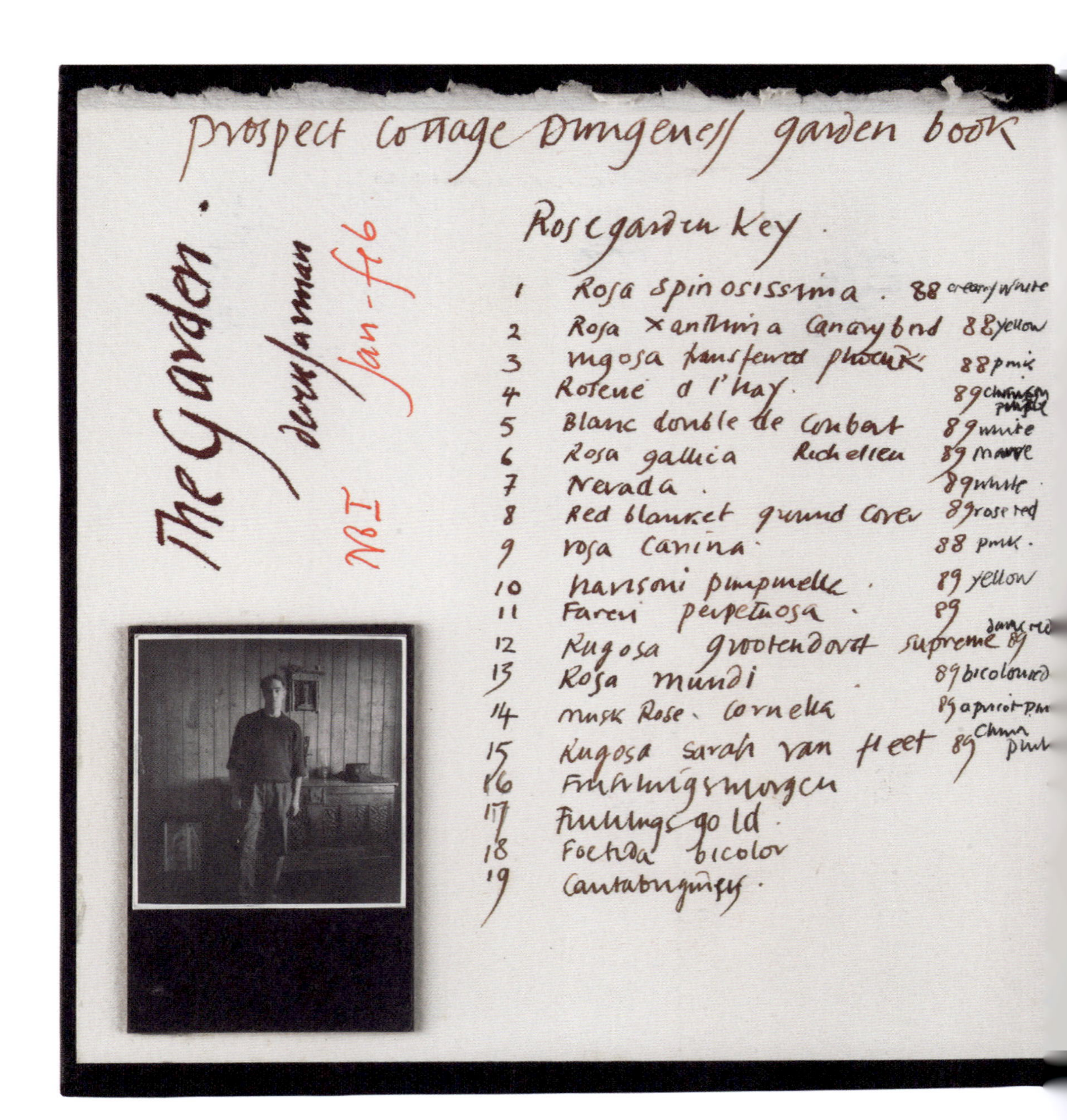

Garden Journal I, January–February 1989

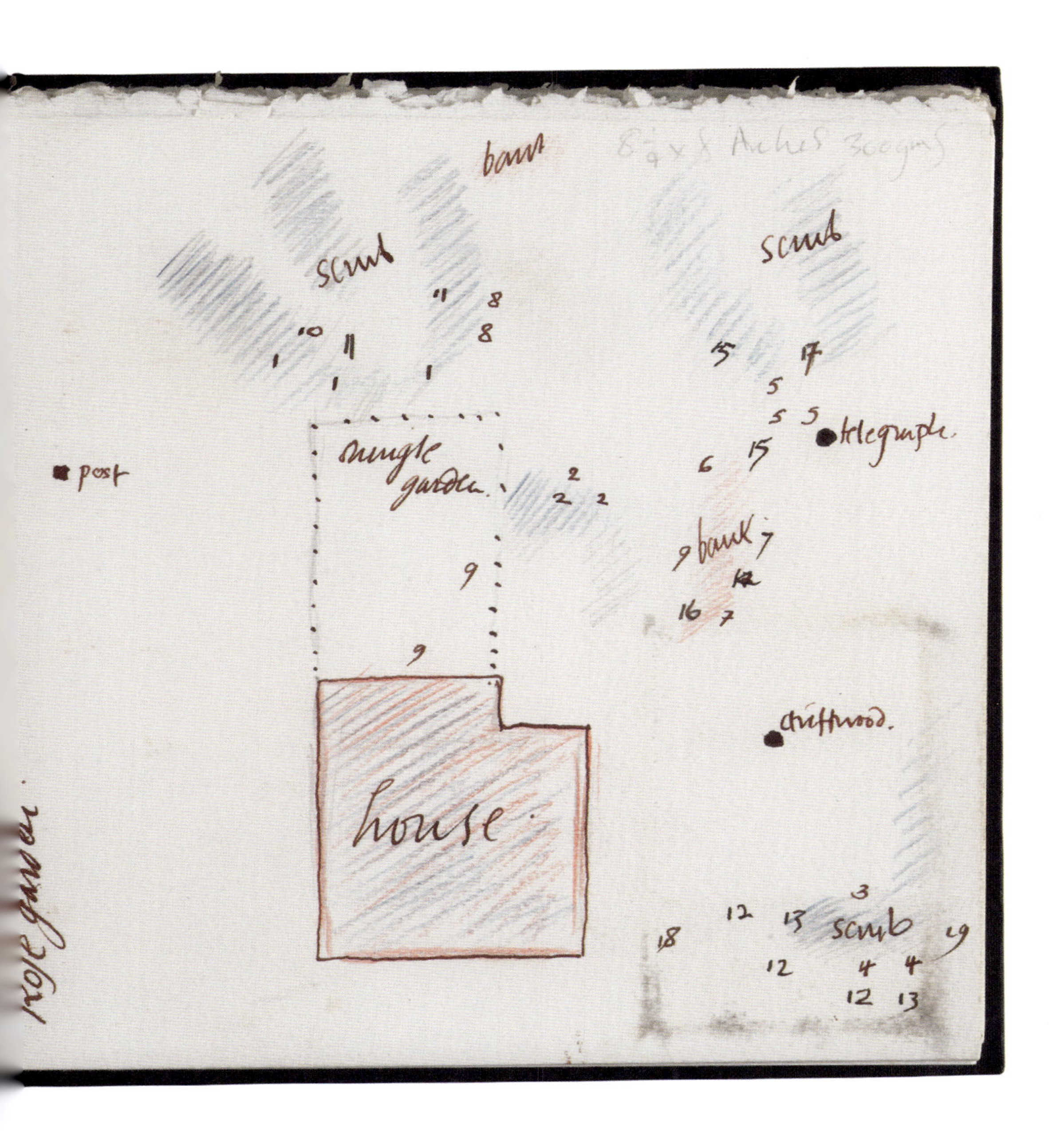
bank
scrub
scrub
post
shingle garden
telegraph
bank
driftwood.
house
scrub
rose garden

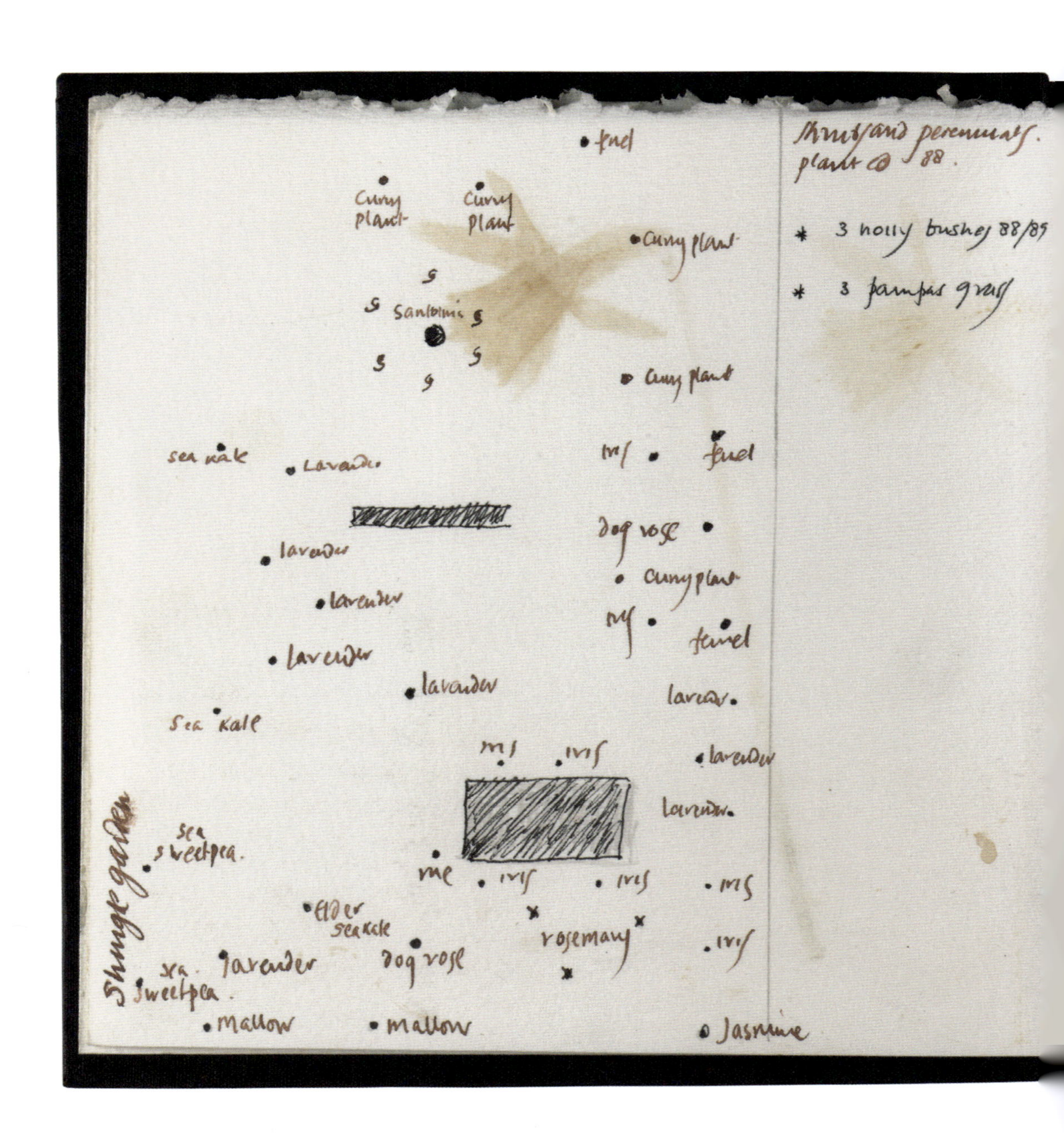

Garden Journal I, January–February 1989

wild flowers.
sea pea
seakale
vipers bugloss
horned yellow poppy
foxglove

wild shrubs –
dog rose.
elder
broom

prologue.

Yesternight I set out from the bright city to the fifth quarter of the globe – here winter is the coldest. The salt sea marsh is wrapped in darkness the ocean gnaws at the shingles without ceasing – wave follows wave like the fleeting years. Spring will be frosty this year – Summer's bright with sunlight far distant. Tonight the heavens burn with innumerable stars white as the bleached petals in the moonlight. I walk in the night throwing long shadows. Cold giants walk beside me. few look for home here

32 I would find spiral ammonites in the rock pools
and it was here that I built my first driftwood
sculptures and photograph them before the tide came in.
Three years ago I returned to Kilve. Stopping in
the new drive in cafe I asked the waitress if
aunt Isobel still lived in Great Beats. She looked
blank. I looked in the phone book she was not
listed. The house was deserted. We drove down the
lane with its views of the sea and Hinkley Point
nuclear power station. past the old mill now a neat
youth hostel. The poplars were also tidied up and
the chantry had become a busy car park. The
whole area has been declared of outstanding
natural beauty. Today I called Telecom to see if
they could find aunt Isobels number. Perhaps if not
in Kilve. Holford? There is no record they said.
The cold wind has fallen. The sea has turned opaque
jade green. I rummage through my books. The
Wordsworth still has some bookmarkers of faded
red and blue papers used for a collage all those years
ago and Dorothys Journal captures it.
'I never saw such a union of earth sky and sea
the clouds beneath our feet spread themselves to the
water, and the clouds of the sky almost joined
them.'

Garden Journal III, April–May 1989

wed april 27

33

I walk in this garden
holding the hands of dead friends
old age came quickly for my frosted generation
cold cold cold they died so silently

Did the forgotten generations scream?
or go full of resignation
quietly protesting innocence
cold cold cold they died so silently

I linked hands at four A.M.
deep under the city
you slept on
never heard the sweet flesh song
cold cold cold they die so silently

I have no words
my shaking hand
cannot express fury
sadness is all I have – no words
cold cold cold you died so silently

Matthew fucked Mark fucked Luke fucked John
who lay in the bed that I lie on
touch fingers again as you sing this song
cold cold cold we die so silently

My gilly flowers roses violets blue
sweet garden of vanished pleasures
come back next year
cold cold cold I die so silently
goodnight boys goodnight Johnny good night Mat
goodnight goodnight

22 Dungeness bathes in a pool of clear sunlight ringed by dark purple thunderclouds - the heat shimmers off the stones there is no wind today - breathless the bees lazy flight through the foxglove spires. my blue columbine is in flower and last years seedlings are thriving. The columbine aquilegia the eagles foot was one of the herbs used against the black death in the 14th century a wild flower that has crept into my garden. The thunder clouds move closer a hawk hovers so high it is nearly invisible. ~~[illegible]~~ here on the stones - blue dragonflies and butterflies are mating. gold cinquefoil and eggs and bacon catch the last rays. the sun is overtaken by the clouds - distant boom of thunder.

Cinquefoil boiled with the fat of children made the witches ointment. spell flower for love potions. its twelve. my noon flower closes shop. Jack go to bed at noon. 'it shutteth itself at ~~noon~~ twelve of the clocke, and sheweth not his face open untill the next dayes sun doth make it flower anew

sat 3 The wind got up and within minutes it was raining so hard it dripped through the roof soon the drive was awash. its ~~was~~ effect on the

Garden Journal IV, May–June 1989

colours in the landscape was immediate as if someone
had brushed varnish across a dull painting. This is
the first rain for over a month – you could hear the
grasses sigh with relief. When the rain cleared ravenous
slugs appeared in their juicy hundreds, to feast
on the poppies and fennel. Lyn arrived from
London swinging her car into the shingles and nearly
burying it. After which we tracked down lucky
stones on the beach for a necklace. Then took off
for a long drive across the marshes to Old Romney
Appledore, Rye, Winchelsea and the cliffs at
Fairlight. Lyn could not believe the farms
thought this only existed in the picture books
the old houses submerged in roses and honeysuckle
we glimpsed peacocks through one garden gate
and a large white owl. We didn't return home
till seven. With books biscuits and fudge –
and two enormous pots to plant geraniums
in front of the porch. By Fairfield church
we gathered elderflower in a lane. Lyn calls
it lace flower. Which we fried in batter and
sprinkled with sugar for supper –

Sun 4 June gilded a small pocket book for my
BLUEPRINT. As I walked along the beach
I thought the film might follow the sound of the
footsteps. A journey with the continuous murmur

Dragonfly, 1986

Song for Dungeness, 1988

Landscape, 1991

Prospect, 1991

Acid Rain, 1992

Oh Zone, 1992

My first photograph of Derek lying on the deck of an old fishing boat, 1990

GARDENING ON BORROWED TIME

HOWARD SOOLEY
GARDENING AND BOTANICAL PHOTOGRAPHER

My first encounter with Derek Jarman was as a young art student, at a screening of the film *The Last of England* (1987) at the Parkway Cinema in Camden Town. The film was inspirational. Back then, in the late 80s, I was unaware of what part Derek and his garden would later come to play in my life.

Time passed and one sunny May morning in 1990, I found myself camera in hand walking across the shingle beach of Dungeness making portraits of Derek, beginning a friendship that would last until the quiet moment of his death.

Commissioned to photograph Derek, I'd left London, driven down the M2 motorway, then onto ever narrowing roads, meandering through the Kent Downs, past coppiced hazel woods, erupting with arums, spilling with green, orchards of setting fruit, and fields of hops laid out on elaborate constructions of tall hazel poles and twine.

As I drove in the direction of the Kent coast, with every passing mile there seemed to be a paring back of the landscape, little by little a stripping away of something deemed unnecessary.

After a while the roads straighten and stretch out into the distance, the trees thin and dwindle, start to contort, formed by the wind. Church spires rise out of the flatness and point up to the sky like thorns, there is an ever growing intensity to the light. You are crossing Romney Marsh. The journey stops feeling transitory, like you are getting from A to B on a map, and more like an odyssey. The road might be a river as you head into a heart of darkness, or you are journeying into Tarkovsky's Zone in *Stalker* (1970). The place names change, there is a darker resonance. On through Old Romney, past St. Clement's church, which seems to roam and graze the marsh like the sheep that surround it (it is here where Derek is now buried along with his partner Keith Collins).

Further on, approaching the town of Lydd the road veers off to the left leaving the old coastline that continues and winds its way towards Rye and Winchelsea, the road now a straight line over a sea of shingle. The light continues to gain in intensity, brighter, cleaner, flooding across the road, filling my eyes, blinding. The green has gone and everywhere is gold, golden shingle, blue sky transected by black crisscrossing power lines radiating out from the outline of Dungeness nuclear power station, rising steadily out of the haze. Pylon after pylon push up like a giant Meccano forest guarding the nuclear power station. The black lines funnel you towards Dungeness.

To avoid driving straight into the sea the road curves sharply round to the east in the direction of Greatstone. Just before the bend is an inauspicious turning on the right where a black painted sign quietly announces you have reached Dungeness, the end of the road and the end of your journey.

As the car approaches the tracks of the Romney, Hythe and Dymchurch Light Railway that run across the road, I am reminded of Cocteau's *Orphée* (1950), crossing the rails of a metaphorical River Styx to enter into the underworld on the other side.

Dungeness ('Denge ness') is a 10 square mile spit of shingle pushing out from the Kent coast into the English Channel towards the French coast. The most south easterly corner of England. A peninsula surrounded by sea and bathed in light. It is a Site of Special Scientific Interest and recognised as an important world geomorphological site with unusual biodiversity, as well as being an important stop for migratory birds.

It is a place like nowhere else; it is indeed the very *Last of England* the ancient 5th quarter of the globe. It is perhaps not just an accident that Derek chose to settle here, make it his home and start a garden.

I am reminded of Jan Morris' eulogy to Trieste in *Trieste and the Meaning of Nowhere* (2001) where she talks about the landscape and city, lovingly and laconically, as a place of exiles. Dungeness has something of this about it. Derek talked of his sense of separateness after a childhood of public schools where he was forbidden to go into the local towns and of living in RAF camps isolated, surrounded by barbed wire, not mixing with the locals.

As you cross over the tracks you leave behind the Kentish Garden of England and you are most definitely somewhere else, through a looking glass, without usual convention and conformity, a place of itself and its unique landscape.

A line of black tar painted fishermen's cottages are strung out along a single track road like a line of broken fairy lights, a lifeboat station, a pub and 2 lighthouses (the first made useless after the building of the power station in 1965 which blocked the reach of its beam) and at the end of the road Dungeness nuclear power station… Derek's 'Emerald City'.

The beach is scattered with the husks of long abandoned wooden fishing boats left behind from the simpler days of inshore fishing, adrift on a frozen motionless sea of shingle, skipper'd by past souls, invisible sails billowing, abandoned by the retreating tide of time. Fishermen's huts nestle in the shingle where they fell from the sky bedecked with rusted tins and tools. From each, narrow gauge tram lines rust their way towards the road where once gluts of herring and pilchard were hauled up the beach in carts to be taken away to market in Rye. Now grass grows and wild peas nestle and snake across the sleepers and vanishing tracks.

The skeletal pyramid that once stood proudly as part of the Pluto pipeline, refueling ships at sea for the Normandy landings, in the mists of time has become a ghost, turned to rust, then dust and now is gone.

Here nothing is permanent, not even the land itself, a temporary shifting spit of shingle, always moving, shaped by the sea or bulldozer.

Back then the pyramid still stood sentinel on the beach. I walked with Derek, squinting in the blinding light, watching his lips mouthing words but unable to hear anything but the gusting wind and crunching of shingle. Everything around us seemed as if it were being eroded, by the sun, sea, the wind, salt and rain. Concrete, rubber, metal, and wood, were all being turned into dust around our feet. All but the flora, which seemed to thrive on adversity.

Stretched out along the shifting shoreline, sea kales filled the air with their heavy sweet scent which carried right the way across the ness.

They were at home pressed up against the sea, glaucous cabbage balls of blue elephant ear leaves randomly dotted throughout the shoreline, one of Mother Nature's more surreal works.

After a storm, when the beach has been washed away, they flail around, anchored to the water table by a rope-like taproot. I have seen them up to 30 feet long. In the autumn gales their dried molecular seed heads snap free and blow like tumbleweed in a Spaghetti Western across the wide expanses of shingle, depositing seeds as they go. It is advised by EDF, who operate the power station, not to eat the sea kale as they store radioactivity, the same is true of the field mushrooms that grow down by the gravel pits.

Prospect Cottage stands roughly midway in the line of fishermen's cottages, a simple child's drawing of a cottage, tar black, two yellow painted window frames for eyes (on either side of the front door), looking out across the beach to watch the sun rise over the sea every morning. Then watch as it tracks and arcs its way unhindered across the sky, eventually skimming the top of the power station, and as a dimming orange ball becomes tangled in the power lines and pylons in the distance towards Lydd.

Prospect Cottage is much as Derek found it the day he visited Dungeness in 1986. Tilda Swinton saw a small for sale sign in the

The concentric lines of shingle that make up Dungeness, photographed from the top of the old lighthouse, 1995

window and in an instant, Derek knew this was the place for him and the purchase was made.

His sketchbooks from then show simple line drawings of a garden plan and a plant list (pp. 22-23), you can sense the intention and excitement. Written in red pen are the words 'gardening on borrowed time'.

Recently diagnosed with HIV there seems something so optimistic and positive about starting a garden in the face of illness. From his earliest childhood he had a strong desire to garden. A desire that had stayed dormant within him, dormant like a seed till that moment. I think Derek knew there was a magic in planting seeds. In no time he had started marking out a garden in the front.

At first glance Dungeness doesn't look like a promising place to start a garden. There is no soil, no beds or borders, no fences or hedges, no shelter from the searing sun or punishing salt laced winds that rip in from the sea. Just shingle, as far as you can see.

In many ways the front of the cottage is a childhood dream of a garden. It was laid out first, initially as a series of geometric shapes drawn with white flints collected from along the beach. A simple white circle whose outer edge was made up from a ring of standing stones, white 'dragon's teeth' flints, planted with a circle of salvias. Then a white square splashed with red and orange poppies like a joyful Fauvist Pollock painting. There is something comforting in the simple geometry as it bridges the chaos between the small cottage and the vast openness of the landscape, order fading into disorder. An inter-zone, where thoughts distil, an artist drawing, a translucent sketch thrown over the landscape, like a welcome mat at the front door.

The planted upright driftwood sticks that have become synonymous with Prospect Cottage owe their existence to the sea kale. Washed up by the night tide, sea worn fossils of another time and place.

Each morning Derek would collect faggots of driftwood from the tide line. The sticks would mark out and protect the delicate purple shoots of the kale, hidden under the shingle like long abandoned landmines, easy to miss in the chaotic pattern of the shingle. He would decorate and adorn them with rusty pieces of ironwork found on walks, often jetsam left behind from the Second World War.

Chance finds of old sea defence groynes would be barrow'd up the beach, inverted and planted in the garden.

The back garden was much more random and fluid from its inception with no tangible framework or structure. Although on the plans there was a notional boundary to the property, without fences, the garden seemingly stretched all the way to the horizon in all directions. The landscape wasn't fenced, any intervention was a dialogue with the whole of the landscape. Something I grew to love and cherish.

The native flora of the ness planted themselves where chance or preference put them. Occasionally Derek would intervene waving dried seed heads over an area that might benefit from a foxglove or yellow horned poppy.

Cotton lavender (*Santolina chamaecyparissus*) and curry plants (*Helichrysum italicum*) were bought from a local nursery up the road in Greatstone. They were at home growing in shingle and kept a compact and elegant form. I still find it astonishing how well the plants would cope in pure shingle. Nursery woman Marina Christopher of Phoenix Perennial Plants in Hampshire, made an experimental pure gravel bed planted with umbellifers, they all seemed to thrive and kept a tighter more architectural form than they did in soil.

Back in those days it was pure shingle on the ness, as far down as you were able to dig. Not even a dusting of soil was to be seen. It is a wonder to me how anything managed to grow at all. Somehow there must have been enough moisture and nutrients held on the surface of the flints to sustain life.

As the years pass the shingle settles and vegetation falls, soil starts to form and establish itself enough to encourage the encroaching grass to sweep and bind its path across the ness, the ness' character changes, one ecosystem becomes another. Back then there was no grass at Prospect Cottage, now like a spring tide it laps at the edges of the garden held at bay only by relentless weeding.

It always felt miraculous to me that anything could grow there, nothing was taken for granted, every plant that found its way, grew and blossomed was special: gardening against the odds. This wasn't wasted on Derek.

Eventually the spaces between the plants started to suggest a path, a way of moving around the garden, and a fluid structure was born. With annuals, bi-annuals and the shorter more intense life cycle of perennials on the ness. This is always changing from one year to the next. The organic nature of the paths and spaces between things became a natural system, like rivulets, becoming a winding twisting beck, free to find its own course. A path devised around the needs of the plants and not the other way round, as is more normal.

Initially most of the plants growing in the garden were native to Dungeness; sea kale, wild peas, viper's bugloss, yellow horned poppy, teasels and sea holly. Others came from the nursery in Greatstone or the wonderful Madrona Nursery in Lydd (now lamentably relocated further afield near Maidstone). Liam at Madrona had a wonderful eclectic collection of plants most of which were unusual and coastal tolerant. His nursery was laid out like an intricate maze shoehorned in a small domestic bungalow's back garden. It was usual to become completely engrossed and disorientated, lost for blissful hours in there, always emerging with something new and untried. The giant sea kale (*Crambe cordifolia*) came from there, which we planted far from the house by an old upturned water tank to give it some wind protection, its two metre plus flower spike would rise into the summer sky like a giant white sail on the horizon.

That summer I spent more and more time on the ness gardening with Derek and less and less time working as a photographer. The preceding few years I'd been working hard, travelling back and forth, but wound up exhausted rather than happy.

Gardening became my cure, a way of slowing down and reconnecting with more important things.

Looking back, there seemed a magic to those days, they seemed timeless and endless (though one day they inevitably ended).

Without hands on the clock, the sun would rise over the wakening sea, golden beams of sunshine streaming into the cottage. With the first creaks of floorboards the day would begin.

I slept on an old hospital bed in what was Derek's painting studio. The bed was a prop from the film *The Garden* (1990) in which Derek lay surrounded by the sea and angels holding flares aloft in the Dungeness sky.

The kitchen was a tiny galley room with only space for two sat on a small bench at old oak table overlooking the back garden. Each day started like this; cornflakes, marmalade, toast and coffee, sat together gazing out of the window at the flowers nodding in the breeze. The sticks and posts being the only vertical structures for miles became welcome perches for the tiny migratory birds navigating by the visual reference point of the ness. Rare Russian warblers and buntings staring curiously in as we stared back over our coffee in wonder.

The first part of the morning would usually be spent in the garden doing various jobs of the day, later a walk along the beach, along the high tide line to see what treasures had been washed up in the night. It's something I love about the sea, every beach is a port for the world's flotsam and jetsam traffic. Bits of orange plastic may be from Mexico or from Mablethorpe. A connection to somewhere else, to mystery, possibility and the unexpected. Twice a day the sea would deposit its treasures on the shoreline. The sticks were amongst the best of them, all curiously around the same two foot length, seemingly ancient, twisted sculptural bones of the sea.

Unknown from which wood, from which tree, and which continent they arrive.

One morning walk with Derek and Peter [Fillingham] after a storm, we found the beach covered in a firmament of fading red starfish fallen from the night sky, suffocating in the growing light. An extraordinary sight, we rushed to throw as many back in the darkness of the sea as we could.

In the afternoons we'd often escape the insular isolation of the ness and venture over the railway tracks to the world beyond. We would visit nurseries further afield, near Rye or Great Dixter or visit bric-a-brac shops like the Aladdin's cave in Appledore stuffed with old gardening tools and flower pots or the second hand book shops of Hastings.

Other nurseries were woven into our different routes to and from London. The A2 via Faversham passing Dr. Tim Ingram's Copton Ash nursery, with his spiky New Zealand Aciphillas. Or down the A20, through Hawkhurst to hellebore heaven at Washfield nursery, guided by extraordinary plantswoman Elizabeth Strangman and Graham Gough as to what treasures to chance on the ness. It was always fun to try something new and to learn something new. But somehow the simpler things in the garden were always the most beautiful. The magenta-headed sea pea snaking over rusted chains, the emerging shoots of kale, purple-black extending alien fingers, unfurling from the golden shingle.

Walking out behind the cottage, towards the outline of the power station, across an English tundra of lichens and mosses, before you reach the old gravel pits, now a thick tangle of willows and RSPB mist nets, jet black lines etch-a-sketch out across the lichen. Stiff pointed spikes, of what appear to be brambles, but if you look again are copses of naturally occurring bonsai blackthorn. An interlocking web of bushes 6 to 10 inches high with a spread of a foot or so across. In February tiny starry white blossom starts to illuminate the darkness of the twigs, in October they are bejewelled with dusty purple black sloes almost impossible to pick with human fingers as there is such little distance between the lancing thorns. After Derek died the meticulous Keith used to painstakingly (and I imagine painfully) pick them for making sloe gin. Once made, the gin soaked flesh of the berries would be dipped in chocolate and were gratefully eaten.

Derek never appeared happier, more at ease and content than when he was gardening. In Jeremy Issac's *Face to Face* interview with Derek (1993), when asked about gardening, Derek declared 'I should have been a gardener' and recalled some of his earliest childhood memories picking red geraniums in the warm Italian summers at Lago Maggiore where his father was seconded after the Second World War. Flitting like a bee from plant to plant nothing else seemingly to matter in the whole world. It's an image of him I hold and cherish, perfect moments and timeless memories. But of course time was ticking, silently or unheard over the buzzing of the bees, the sound of the waves or the mournful lament of the fog horn in the mist.

Throughout all the time I knew Derek and spent time with him, illness was always there. AIDS in all its forms, its multitude of masks, always lurking, the unwelcome visitor who came to stay. You could slip from it for a moment, for a morning but it was always there, waiting in the darkness. Blinding, drug-addicted, insatiable, unrelenting and malevolent.

Derek fought his illness with openness, busying himself with so many things that there was little time or room left for illness. In a way it worked, and for a time he cheated death hiding amongst the flowers and dancing with the bees. The love and friendships, the caring and sharing over those last years defied the background they were set against.

Derek thought that although no one had lived through AIDS, one day someone would, why not him? To the last he was always open to being the first.

Chemists and doctors struggled against the sweeping hands of the clock to find a drug or combination of drugs to stop AIDS. Everywhere people were dying, beautiful young lives being extinguished, a meaningless slaughter of youth. The Andrewes ward (a specialist AIDS ward) at St Bartholomew's Hospital like a battlefield of the falling and felled behind every door, every curtain, hunched over in a corridor the dead and dying. No one walked out of there alive. You had better odds at the Somme or on a passenger airliner slamming into a mountain side. Then, nobody survived AIDS.

On the wards there was an absolute absence of hope, the hopeless crawling into their graves. Families and friends like distraught parent birds round an empty nest. All hope died there and I tremble as the memories flood back like bad dreams.

Derek's visits to the Andrewes ward became more frequent and our days on the ness correspondingly fewer. He would spend his days writing from a hospital bed, time starting to race, chasing his pen across the page. The ageing process of his body quickening and spiralling out of control.

Once or twice a week he'd break out of Barts, disconnect from the drip, put on some by now ill-fitting clothes, and be helped into my waiting car. I'd drive through the city down into the darkness of the Blackwall Tunnel, spilling out onto the A2 and back onto the rolling downs past hazels and orchards. To where the birdsong drowns out the chirp of the drip.

Frail, becoming blind, skin on fire with a rash, he'd open the door of the cottage, picking up the mail in the hallway, taking off the ill fitting hospital clothes, climbing inside a huge djellaba. Straight into the garden, happy till the last light of the day, until the calling of the drip could no longer be ignored. Each time leaving never knowing if that was the last time.

There was an inevitability to the end, it could no longer be ignored. A place to be buried, a headstone carved. One day in Barts we went for a walk, ending up in the hospital chapel. Empty and silent we

Derek gardening in his djellaba on a day trip from St. Bartholomew's hospital, October 1993

sat at the back like reluctant school boys. A cold hard pew in the January darkness, leafing through a worn musty hymn book to find something suitable for his funeral service. With the church's past record on homosexuality, it didn't feel that promising, more like depressing, forced into playing by establishment rules, becoming an outsider, again. It was painful to be there. Cold, still, dark and dank reading aloud verses to each other and trailing off to silence at the sight of the shepherd's crook. I remember looking up, I see it so clearly, even now, the colourless light, silent tears slowly rolling down Derek's cheeks. I'd never seen him cry before or after. He started reading aloud 'All things bright and beautiful' a lament for things to be lost.

I remember back to that first May morning, walking over the shingle alongside Derek. The first photograph I made of him (p. 34). He was lying serenely on his back in a long abandoned fishing boat, eyes closed, face turned to the sky, bathed in light. It always seemed like a rehearsal for that last day, drifting silently away on a morphine breeze, from the shores of the Andrewes ward, one misty February morning in 1994. The first and last point in our circle.

Coming back from our allotment today after planting the broad beans and boarding the bus in Hampstead, a recorded voice announced, as it does every time 'the 46 to St. Bartholomew's hospital' and I feel a chill.

Following pages: Images taken by photographer Howard Sooley at Prospect Cottage on his visits to Derek Jarman between 1990 and 1994.

PROSPECT COTTAGE

THE JARMAN GARDEN EXPERIENCE

CHRISTOPHER LLOYD
GARDENING CORRESPONDENT,
COUNTRY LIFE AND THE GUARDIAN

It was on 24 June 1990 that, quite accidentally, I happened on Derek Jarman's garden. Five of us - Beth Chatto being the picnic maker - plus my two dachshunds, spent this sunny day on the beach at Dungeness. The shingle drops steeply into deep sea and Channel traffic passes quite close by. No bathers - too dangerous for that, and there have been numerous shipwrecks here, where opposing currents meet. There are anglers and plenty else to watch. Robin Barrett constructed a little boat from bits of flotsam lying around, using a scrap of stiff plastic for a sail. He launched it and we watched it bobbing merrily about till it disappeared from sight in the direction of Hythe.

All the land behind us has been wrested from the sea, shingle piling on shingle - beach, it is known as, locally - a huge, ever-increasing promontory of it. Presently, a fishing boat came in and landed its catch a couple of hundred yards along the shore. Leaving the others to carry our things to my car and to drive it closer to the fishing boat along the road that runs parallel to the shore, Beth and I hastened to see what had been caught. Nothing exciting (no John Dory, which I had first seen here many years before). The men were grumpy, anyway; tired, most likely.

Dachshund and Christopher Lloyd on Dungeness beach, 1990

Previous page: *Derek Jarman at Prospect Cottage in the snow*

We walked towards the car and then I spotted some brilliant flower colour; I headed for it. Houses, well separated, are strung out along this road on its inland side. There are no visible boundaries or fences. I made a beeline for that colour. It hadn't just happened, like the wild flowers that make such vivid displays beneath the dazzling openness of the sky, yet nowhere else were any of the habitations making the slightest attempt to garden. Here, around a timber cottage tarred black but with cheerful yellow window frames, a garden had been made.

Most brilliant were the Californian poppies (*Eschscholzias californica*), wide open to the sun. But so much else. 'Come over here, Beth', I called out and a moment later we were both excitedly examining the apparently deserted scene, myself taking photographs and scribbling a list of plants so evidently happy there, in the back of my pocket diary.

Presently, from a neighbouring house, a young man came up behind me, as I crouched. 'I expect you want to know what I'm doing', I said to him, 'I can see what you're doing', he replied. And he left us to it.

June is a good moment in this area. Gale-force winds have temporarily let up and there is a great burst of colour from wild flowers, many of which were included in this garden, either because they were already there or because they had been added. But although the vegetation of early summer had masked a good deal, there was evidence of design here, with arrangements, often found in circles, of flints and other stones, sea-smoothed bricks, sea shells and more besides. Also of 'sculpture' with all kinds of legacies from the war, fishing accessories and sea defences.

We gradually worked our way around to the front of the house where two raised beds with a fair amount of soil in them were growing a different range of plants from those which the shingle could support. How surprised we were when the door opened and Derek Jarman stepped out. My young friends knew him at once and were star-struck. Being of an older generation and having ceased to visit movies, I did not know him and he did not expect me to. It was not long afterwards that I made a point of seeing his latest film, *The Garden*, in London, and I then realised what all the interest and excitement was about.

Jarman knew about Beth, through a friend; of her garden and nursery. Whether he had heard of me, I do not know, but he visited my nursery fairly frequently towards the end of his life and got to know my garden, which he mentioned very appreciatively in his own garden book. I came to Prospect Cottage again, two summers later, and met him once at my home, Great Dixter. He was frail and in a wheelchair. I had hoped to get him over to dinner, but that never happened.

His garden lies around the house, but mostly fore and aft in roughly equal proportions. The principal feature in a flat landscape is the Dungeness nuclear power station. It is not a romantic scene, being scattered with huts and sporadic habitations, with pylons and overhead cables and power lines. But it is open and smells good. Mainly of the sea, of course, but also of flowers and plants in their season.

Sea kale (*Crambe maritima*) is the dominant plant, more abundant here than anywhere else in the British Isles. It is the first colonizer above the tideline, with taproots that delve in search of moisture. Jarman quoted 20 feet as a root length, measured on a plant that had become exposed after a storm. If you were to plant sea kale, you would need to start with a section of root, and that would make shoots and its own roots. But I fancy the plants were all in place in this garden already; Jarman made them the centre of formal circles that he built.

Young sea kale shoots first become visible in March and the foliage is purple, later becoming blue-grey. It is crinkly and cabbage-like, the flowers are gathered into large, white cumulus clouds that smell of honey on the air, the first to open, in early May being those that are lying against the 'beach' and receive extra heat from the sun-warmed stones. The flowers are followed by huge tangles of pea-sized fruits, which take months to ripen and eventually become 'the colour of bone. At this stage they are at their most beautiful - sprays of pale ochre, several thousand seeds on each plant' (*Derek Jarman's Garden*, p.16).

Gorse (*Ulex europaeus*) is abundant on the ness and sends the smell of coconut into the air. Rabbits graze gorse as high as they can reach, which makes it very dense. When flowering, the rabbit-grazed surfaces are solid with blossom. Jarman planted two circles of gorse, with a baulk of timber standing vertically in the centre of each. Blackthorn blossom (*Prunnus psinosa*) is a great feature, all over the shingle in April, the bushes far lower than normal, often hugging the ground and flowering early as a consequence. Jarman waged war on the browntail moth caterpillars that wreak havoc in blackthorn foliage.

Broom, *Cytisus scoparius*, comes next again reduced in height and making pools of yellow, of a brighter shade than gorse but sour-smelling. Other good scents, mainly from the garden, include santolina, the grey lavender cotton, planted in formal circles. To be kept neat, it needs clipping and this prevents flowering, which is an undistinguished event anyway. Then, *Helichrysum angustifolium*, the curry plant which smells strongly of exactly that. With narrow, grey leaves, it is used formally like the santolina, but is allowed to carry its huge crop of yellow flowers, being clipped over in August. 'Helichrysum, in which the lizards dance, is the backbone of the garden, both in the formal garden at the front and at the back', Jarman wrote.

Herbs are strongly featured. Jarman describes *Ruta gravolens* as turning 'into a tight rue football'. He probably gave that a hard, annual clip-back. Lavender is very much at home. Any plant with a hairy, grey-white leaf, or glaucous, with a wax-coated leaf (like sea-kale and rue), is geared to coping with the sun, wind and arid conditions at the root. Jarman was always a passionate gardener, from earliest childhood, and he well understood which plants would like him under these extreme conditions. On my heavy clay, lavender hates me, growing scraggily and frequently leaving me with gaps in lavender hedging. I have learned to live without it. Rosemary, closely related to lavender, thrives at Prospect Cottage, as do sage and marjoram. The shrubby, grey filigree-leaved old man, *Artemisia abrotanum*, with its curious, pungent smell is here, and also absinth, *A. absinthium*. Lovage and fennel have deep, fleshy roots and they are good. Borage grows here but its cousin, the biennial viper's bugloss (*Echium vulgare*), is one of the showiest natives. In its first year, it makes a starfish of ground-hugging foliage. In its second, it sends spikes of blue flowers up to 2 feet high. Imagine them combined, as here, with crimson field poppies, *Papaver rhoeas*. But Jarman also loved 'the sky blue cornflower which comes up in every corner'. That is an excellent poppy companion.

The wild, or introduced and feral, plants on the ness are a significant feature. Some were already in this garden. Others Jarman introduced so as to have them in a concentration, which is what gardening is largely about. He did not have to bring them far. In my garden, I consider toadflax (*Linaria vulgaris*) a weed, not merely because it has a running, invasive rootstock, but because it grows in a straggling, weedy way. Not here. Colonising, yes, but short and with a dense spike in two shades of yellow - a kind of mini-snapdragon with a spur at the back. We normally think of foxgloves as tall inhabitants of coppiced woodland. Here, they are equally happy but not much more than 2 feet high and with more intensely rose-purple flowers on an exceptionally dense spike.

The horned poppy, *Glaucium flavum*, has scalloped grey foliage in its first year, which alone would make it worth growing, but in its second makes a 2- to 3- foot branching inflorescence, producing a succession of fragile yellow poppy flowers, each lasting just one day. Valerian (*Centranthus ruber*), pink, red, and white, grows near the road over many parts of the ness. Jarman rightly observes that it will have a second flowering, particularly if cut back after the first flowering in June. 'I have always loved this plant. It clung to the old walls of the manor at Curry Mallet which my father rented in the early fifties, and grew in the garden of the bomb-damaged house at the end of the road which the airman Johnny, my first love, took me to on his motorbike, with my hands in his trouser pockets - so valerian is a sexy plant for me.'

Rest-harrow, *Ononis repens*, was a favourite and Jarman showed me where it was growing on the roadside nearby, better than in his own garden. It is a spiny, grey-leaved plant that makes a foot-tall mound and is covered with pale pink pea flowers. Normally, I associate it with chalk downland, but alkaline conditions prevail equally along our shores, where broken seashells abound.

Jarman grew not only *Rosa rugose*, which is introduced but wonderful near the sea, and the dog rose, *R. canina*, but also the burnet rose, *R. pimpinellifolia*, a true wilding of the shingle. Only a foot tall, under starvation conditions, it creeps around making a colony and opens its cream-coloured flowers as early as May.

Elder is one of the taller shrubs, here, but still not much above 6 feet high, as its wood is soft and anything with the temerity to put its head up must take a battering. I think it looks dreadful under such conditions, but Jarman loved it. He had four specimens, one at the front, three at the back. He valued them for the culinary uses that their flowers can be put to, quite rightly ignoring their fruits, which taste disgusting. But in respect of the shrub itself, he saw a beautiful shape 'made of a thousand small branches'. The fact that the plant kept its 'bone-coloured dead branches' made it look the more attractive to him, as did the dead branches of dog roses, 'pink-grey, swaying in the wind'. This bleached, dead look strongly appealed to him, witness the abundance of driftwood that was brought in to ornament the garden.

Two large, loose shrubs, soft in texture and fast growing, that flourish by the sea are the pinky-mauve tree mallow, *Lavatera olbia*, and the so-called tree lupin, *Lupinus arboreus*, with pale yellow, sweetly scented spikes and fresh green foliage. It always surprises me that such soft shrubs, breaking up so easily when gale-force winds blow in inland gardens, yet should never be so happy as on the coast. The mallow flowered next to Derek Jarman's 'throne', hard up against the inland side of the house. Here, too, grew the exotic relative of sea kale, *Crambe cordifolia*, which makes an 8-foot-high cloud of tiny white blossom. Much more exposed, the giant cardoon stands strong against all winds. Drawn up to 9 feet in my mixed border, it needs the stoutest staking. This is the wild version of globe artichoke, with smaller, more numerous heads and much more prickly.

I felt sorry for bearded irises, whose fragile blooms get torn immediately on opening, even in quite moderate winds - as winds go, hereabouts. Tougher stalwarts included yucca and phormium, the New Zealand flax, often planted as a first defence against the wind, along coastal shores.

Of smaller, ground-hugging plants, the silver cushions of pinks and carnations are in their element. And thrift, *Armeria maritima*, the sea pink of Scotland, but its cushions are green. Jarman actually bought his first plant of this but it soon became fifty and was poised to leap across the road. Because it did not come of its own volition, I suppose ecologists everywhere would disapprove. The sea campion, with inflated bladders behind white petals, is everywhere on the ness and is in this garden. Two native stonecrops make pools of fleshy leaves, tight against the stones. *Sedum acre*, a dashing shade of yellow; *S. anglicum* with flesh pink stars. Gazanias hail from South Africa, but revel in conditions such as these, their daisies, in many shades and so beautifully marked, expanding to the warmth of sun and shingle. Seldom hardy, inland, their chances of being perennial would be more than doubled here.

Jarman did not leave matters to chance when seeking to establish new plants. To make a planting hole, the shingle had to be excavated and half of it would always roll back, so it was an effortful job. Then he put in well rotted manure from the farm up the road, and the stones were returned at the end of the operation. He told me that he dribbled the seeds of annuals into the cracks between stones. If one took off, thereafter it would ripen and disseminate its own seed. That was the way with eschscholzias, wallflowers, stocks and marigolds (*Calendula* not *Tagetes*).

There is a vegetable and herb garden. The bed was lined with black polythene, to keep in the moisture - this spot has the lowest rainfall in Britain - filled in with topsoil, delivered by dealer, and then well dunged. The vegetables and herbs protect the beehive. Pounds of honey were made, far more than Jarman wanted for himself, so some was given away and plenty left for the bees. Flowers on the ness provide abundant nectar. The nearest rape, four miles distant, would be beyond the bees' normal foraging range.

The garden's layout and ornamentation make a very strong impression. The area between road and cottage, down which a straight driveway once led, now consists of a series of large circles outlined with stones which Jarman collected himself, a few at a time, arranged according to colour, the elongated ones on end. Flints can be large and long (high), and could form the circle's hub, though sea kale also makes the focus, in some instances. Shells and coloured stones from the beach make up the surface between circles of standing stones or dragon's teeth.

Any stones with holes in them have been collected and threaded into necklaces, which are hung in all sorts of places, or dropped over anything pointed, like the tines of a garden fork. Sculptures, which are mostly found objects, sometimes assembled, sometimes standing on their own, consist of anything that the area can produce, including the heads of old garden tools. On the morning of my second visit, a schoolgirl had brought a piece of twisted iron which she had found. The balls of metal floats are here and the corks which are another kind of float. Chains, anchors, a hook, wartime fence posts, with one end in a spiral for the threading of barbed wire, and much more besides. I never saw this scene in winter, when it must have looked stark, but in summer, all was softened by vegetation.

This was and still is a quite extraordinary garden. Derek Jarman made and retained devoted friends, who helped him increasingly as he became feebler. Good luck to them, in keeping Prospect Cottage and its garden going. They are besieged with visitors, swarming to this Mecca. Many of my own visitors make a point of going to have a look. Prospect Cottage is very much in the English garden tradition, showing a love of plants and growing them well as a personal satisfaction. Like all good gardeners, Jarman worked with the natural conditions presented to him by the locale. The fact that he was not put off by being told that gardening at Dungeness was 'impossible', and of there being no protocol or guidance, is also in the English tradition. I feel privileged to have been on the scene when I was, though sad that an acquaintanceship had not the time to become much more than that. Jarman was a man whom I regarded deeply and his garden was a manifestation of great depth and total originality.

First published in *Derek Jarman: A Portrait*, Thames & Hudson, 1996

Derek Jarman at Prospect Cottage, c. 1989–1990

This is where the west wing will be built

WHERE FRAGILE PLANTS MAY BE MICROWAVED

ANNA PAVORD
GARDENING CORRESPONDENT, THE INDEPENDENT AND AUTHOR

You could scarcely find a crazier place to garden than Derek Jarman's place, clinging to the great shingle bank of Dungeness on the south coast. Behind him is a nuclear power station, in front of the English Channel, all around the sort of wind that seaside publicity officers call bracing. There is not a single tree or a scrap of earth in sight.

The tiny wooden cottage, tar-black with yellow windows, could have been built for a toy-town set. It is the sort of house a child draws, low, symmetrical, a door in the middle, a window either side. It sits, washed up on the stones, with no boundaries or fences, surrounded by a garden as bizarre as the house is straight.

The front garden seems to have grown up as though, instead of sowing seed, Jarman had scattered different sorts of pebbles and stones. Flat, formal shapes, circles, squares and rectangles are marked out in coloured pebbles.

There is a big square of white stone bordered with weird, rounded flints, and a series of concentric circles in different colours and sizes neatly finished off with fat bushes of cotton and lavender. One circle is laid out like a wheel, the spokes of bleached driftwood, upright stones in the centre angles.

Some plants, such as the grey cabbage-leaf seakale, covered in huge heads of sweet-smelling white blossom, are part of a wild landscape. Others have been introduced - the feathery Californian poppies poking up between the stones, for instance.

The front garden - which works horizontally - seems to trace some forgotten allegory. The back is a vertical garden, a phallic forest of tall thin pieces of driftwood, stuck upright and crowned with hollow pebbles, fishing weights, rusty springs and other bits of ironmongery.

The driftwood, other angular sculptures and the standing stones seem to be growing from the garden in the same way as the plants, flotsam among the flowers. There is a big square iron tank filled with small clay pots sunk to their rims in earth. Each has a different plant in it; wallflowers and woad. Outsized, necklaces of an anchor chain surround the mulleins, foxglove and iris planted in the shingle. The colours are washed-out - the driftwood, silvery grey, many of the garden plants silvery grey, too.

Artemisias - or beach wormwood - santolinas, lavenders and sages cope best with the extreme conditions on the beach. Wind is a problem. Salt is a problem. But the worst problem is summer sun, for the shingle sops up the heat and microwaves any plants that are not used to it. Trial and error have sorted out the survivors. Some of the best are the native plants of the shingle, the kale, yellow-horned poppy, white bladder campion, blackthorn sculpted by the wind into low mounds.

Parsley, sage, fennel and lovage all thrive, shooting up between the stones with astonishing lushness. Irises have been another unexpected success, roses are struggling. Scotch briars, Rugosas and a pale-yellow 'Canary Bird' rose mould themselves grimly to the contours of the land.

Derek Jarman, however, is more entranced by the one flower on his 'Canary Bird' than he would be by a whole bed of roses in bloom further inland in the lush gardens of Kent and Sussex. 'I *like* the feeling of it being almost impossible,' he says. 'It's what I like most in this garden. Every flower is a triumph. I've had more fun from this place than I've had with anything else in my life. I should have been a gardener,' he asserts, as though he had been wasting his time writing, painting and making films.

'Oh, gardening is *much* better than filming,' he says. 'Filming was fun at the beginning, because none of us believed in it or took it too seriously. There's very little excitement in it for me any more. There are too many other people involved in it.'

He has filmed the garden but is not particularly pleased with the results. 'It's something about the light. It just doesn't work on film. It's changing all the time, reflecting off the sea and the pebbles and the silver leaves of the plants.'

He might pass for a gardener, if cast as one in somebody else's film. His face is weathered, an outdoor sort of face.

Gardening is a relatively new passion for him. It started when he bought the beach cottage at Dungeness. He discovered the area when he was looking for a location for a pop video (he made some memorable ones for the Pet Shop Boys).

'I bought the cottage as a joke. That was in 1986. It was just such a weird place, a sort of wasteland with that great nuclear power station and the gravel pits. But the foreshore was marvellous, old fishing boats, huts, rotting winches and, of course, all these extraordinary wild plants.'

When he bought the place, it had a bit of a drive over the shingle to the front door and two small rectangular rockeries made from bricks and broken concrete either side of the front door. After each walk on the beach, Jarman returned to his cottage with more and more bits that had caught his fancy: strangely shaped stones, ancient bits of timber pickled to the consistency of stone, hoops from brandy kegs.

Collecting bits is not unusual. Putting them to some use is. Jarman's garden, drifting indistinguishably into the beach around it, eclectic but triumphantly successful, is one of the strangest things you will see outside your dreams but because it has been made with things borrowed from the landscape around it, it has a curious sense of fitness.

There is a well tuned eye at work here, the same eye that makes the films and that once designed stage sets. Walking around, tossing encouraging words to plants spread-eagled on their noses in the gale, Derek Jarman sums up his creation: 'Yes, I think the back's finished now,' he says. 'There's a danger of putting too much in. It's best left now.'

The front is more liquid, the patterns changing as new trophies come back from the beach, new centre pieces for yet more arcane designs.

'Everybody said nothing would grow,' says the film-maker who has thoroughly enjoyed putting the Jeremiahs wrong. When a new plant goes in he helps it on its way with a couple of shovel-fulls of well rotted muck pushed in underneath the shingle. After that, it is on its own.

He loses about half of everything he plants. The self-seeders, such as the aquilegias, are the most successful, and those that naturally crouch low to the ground (thrift, pinks). Small tepees of driftwood lashed at the top with string protect new young plants from wind, cats and rabbits. Yuccas get windburn. Red hot pokers look completely at home. Peonies are impossible. This is his only regret.

First published in *The Independent on Sunday*

Following page: Edina van der Wyck, *Derek Jarman on Dungeness beach*

FE18